Centerville Library
Washington-Centerville Public Library
Centerville, Ohio

LET'S PLAY SPORTS!
Fishing

by Kieran Downs

BLASTOFF! READERS 2

BELLWETHER MEDIA • MINNEAPOLIS, MN

Blastoff! Readers are carefully developed by literacy experts to build reading stamina and move students toward fluency by combining standards-based content with developmentally appropriate text.

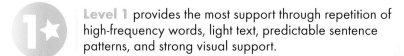

Level 1 provides the most support through repetition of high-frequency words, light text, predictable sentence patterns, and strong visual support.

Level 2 offers early readers a bit more challenge through varied sentences, increased text load, and text-supportive special features.

Level 3 advances early-fluent readers toward fluency through increased text load, less reliance on photos, advancing concepts, longer sentences, and more complex special features.

★ **Blastoff! Universe**

This edition first published in 2021 by Bellwether Media, Inc.

No part of this publication may be reproduced in whole or in part without written permission of the publisher. For information regarding permission, write to Bellwether Media, Inc., Attention: Permissions Department, 6012 Blue Circle Drive, Minnetonka, MN 55343.

Library of Congress Cataloging-in-Publication Data

Names: Downs, Kieran, author.
Title: Fishing / by Kieran Downs.
Description: Minneapolis, MN : Bellwether Media, Inc., [2021] | Series: Blastoff! readers: Let's play sports! | Includes bibliographical references and index. | Audience: Ages 5-8 | Audience: Grades K-1 | Summary: "Relevant images match informative text in this introduction to fishing. Intended for students in kindergarten through third grade"– Provided by publisher.
Identifiers: LCCN 2019054185 (print) | LCCN 2019054186 (ebook) | ISBN 9781644872161 (library binding) | ISBN 9781618919748 (ebook)
Subjects: LCSH: Fishing–Juvenile literature.
Classification: LCC SH445 .D69 2021 (print) | LCC SH445 (ebook) | DDC 639.2–dc23
LC record available at https://lccn.loc.gov/2019054185
LC ebook record available at https://lccn.loc.gov/2019054186

Text copyright © 2021 by Bellwether Media, Inc. BLASTOFF! READERS and associated logos are trademarks and/or registered trademarks of Bellwether Media, Inc.

Editor: Christina Leaf   Designer: Josh Brink

Printed in the United States of America, North Mankato, MN.

Table of Contents

What Is Fishing?	4
How Do People Fish?	8
Fishing Gear	16
Glossary	22
To Learn More	23
Index	24

What Is Fishing?

Fishing is the sport of catching fish. It is one of the most popular outdoor activities in the world!

In the United States, bass and crappies are favorite fish to catch.

largemouth bass

Tournaments allow **anglers** to catch fish by themselves or in teams.

anglers

Champion Spotlight
Jacob Wheeler

- pro angler
- Major League Fishing (MLF) Bass Pro Tour competitor
- Accomplishments:
  - 4 MLF wins
  - 2019 MLF World Champion

Whoever catches the highest total weight of fish wins!

How Do People Fish?

Anglers start by **casting** their **lines**.

They wait until a fish bites. Then they **reel** in their catch.

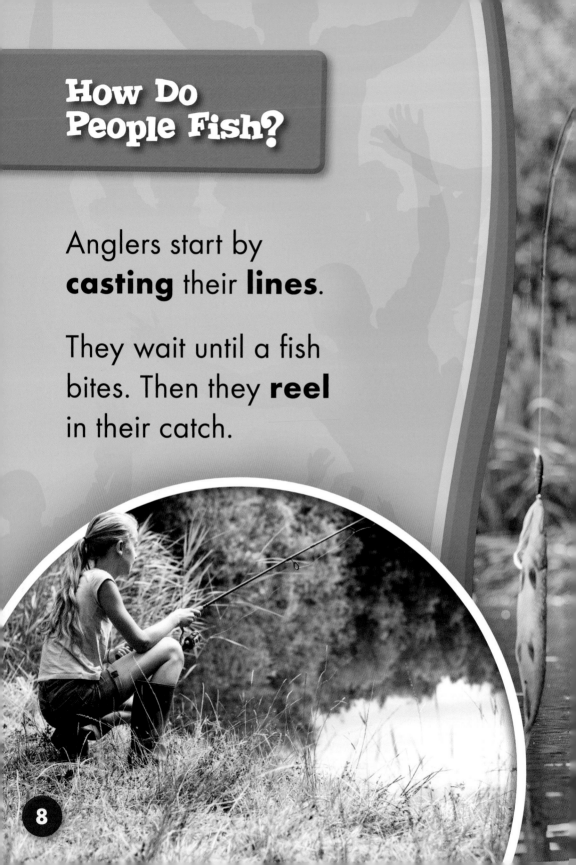

Some anglers keep their fish to eat.

Others **catch and release**. They let the fish go back into the water.

Fishing has laws to keep fish numbers up. Some fish can only be caught in certain times of year.

Usually people can only keep a certain number of fish.

In tournaments, anglers score points for catching fish.

Anglers move on based on score. One person wins in the end.

Fishing Gear

rod

Anglers need a lot of gear to catch fish. **Rods** let them hold fish on their lines.

Reels let anglers bring their fish in.

reel

Bait gets fish to bite the **hook**. It is usually things that fish eat.

Lures also get fish to bite. **Tackle boxes** keep gear in one place.

Bobbers keep hooks at a certain **depth**. They dip down when fish bite.

Some anglers fish on boats. Others stay on shore. They try to catch a big one!

bobber

Glossary

anglers—people who fish with hooks and lines

bait—something used to get fish to bite a hook; bait is usually something fish eat such as minnows or worms.

bobbers—small, floating pieces attached to fishing line to keep the hook at a certain depth

casting—throwing a lure or bait on a line out over the water to catch fish

catch and release—to let a fish go back into the water after it is caught

depth—a distance below the water

hook—a curved piece of metal that fish bite in order to be caught

lines—long threads attached to hooks used to pull in fish

lures—fake bait; lures look and move like something a fish would eat.

reel—to bring in a line by turning it around a wheel-shaped object called a reel

rods—thin, straight poles with reels and lines on them used for fishing

tackle boxes—boxes used by anglers to store their gear

tournaments—contests in which winners continue to play until only one person or team is left

To Learn More

AT THE LIBRARY
Maas, Dave. *Kids' Guide to Fishing: The Young Angler's Guide to Catching More and Bigger Fish.* Mission Viejo, Calif.: The Quarto Group, 2018.

Morey, Allan. *Freshwater Fishing.* Mankato, Minn.: Amicus Ink, 2017.

Palmer, Andrea. *We're Going Freshwater Fishing.* New York, N.Y.: PowerKids Press, 2017.

ON THE WEB

FACTSURFER

Factsurfer.com gives you a safe, fun way to find more information.

1. Go to www.factsurfer.com.

2. Enter "fishing" into the search box and click 🔍.

3. Select your book cover to see a list of related content.

Index

anglers, 6, 8, 10, 14, 15, 16, 17, 20
bait, 18
bass, 4
bite, 8, 18, 19, 20
boat, 15, 20
bobbers, 20
casting, 8
catch, 4, 6, 7, 8, 11, 12, 14, 16, 20
catch and release, 11
crappies, 4
eat, 10, 18
gear, 16, 19
hook, 18, 20
laws, 12
lines, 8, 9, 16
lures, 18, 19
points, 14
reel, 8, 17
rods, 16
shore, 20
tackle boxes, 19
teams, 6
tournaments, 6, 14
United States, 4
water, 11
weight, 7
Wheeler, Jacob, 7

The images in this book are reproduced through the courtesy of: Annabelle Breakey/ Getty Images, front cover (boy); Kris Wiktor, front cover (background); Jeff Feverston, p. 4; Robert Daly, p. 5; Artyon Geodakyan/ Getty Images, p. 6; ZUMA Press/ Alamy, p. 7; patat, p. 8; Presslab, p. 9; Chris Stein/ Getty Images, p. 10; goodluz, p. 11; oneSHUTTER oneMEMORY, p. 12; Dan Thornberg, p. 13; Cal Sport Media/ Alamy, p. 14; okenemada1, p. 15; Iakov Filimonov, p. 16; EvgeniiAnd, p. 17; Dobo Kristian, pp. 18-19; Kateryna Dyellalova, p. 19; kulikovv, p. 20; MaxTopchii, p. 21; teka12, p. 23.